HOW DO PLANTS GROW?

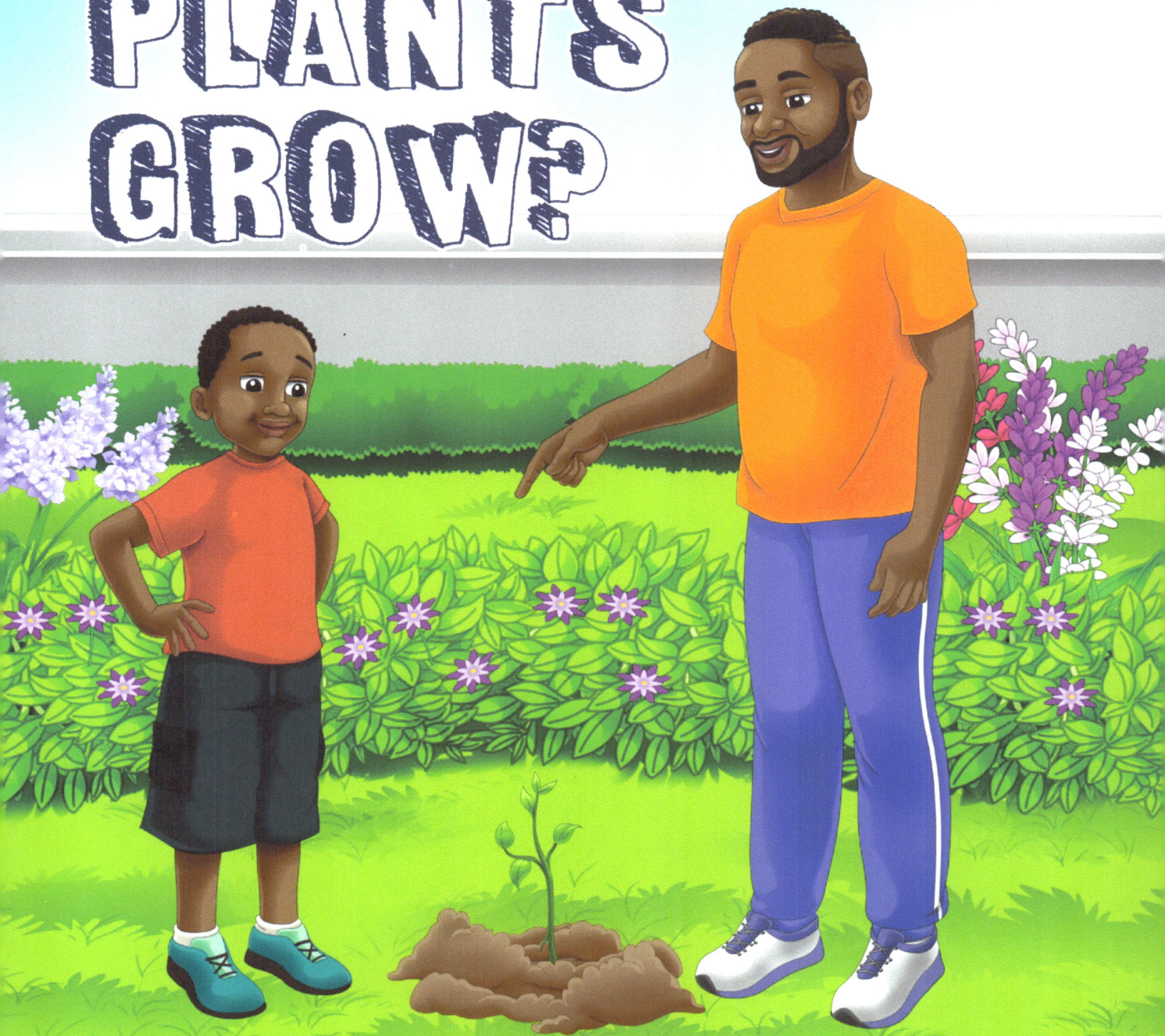

G.A. Sealy

I0073626

Copyright © 2017 G.A. Sealy
DaWit Publishing LLC

ALL RIGHTS RESERVED

No part of this book may be reproduced or transmitted in any form whatsoever, electronic, or mechanical, including photocopying, recording, or by any informational storage or retrieval without the expressed written consent of the publisher, except in the brief quotation in critical articles or reviews.

First Edition (Large Print)
Printed in the U.S.A.

978-0-9965978-8-3   Large Print Paperback
978-0-9965978-9-0   Large Print Hardcover

Library of Congress Control Number: 2017913938

Books may be purchased by contacting the publisher and author at dawitpublishing@gmail.com or visit www.dawitpublishing.com

Acknowledgements

I would like to thank Juju, and the rest of my family and friends for all of their support, and Dr. Marcus Broadhead who helped me edit my book in its early stages.

One day Kelvin and Dad were taking a walk. They were looking at the trees and flowers in the park.

When Kelvin said, "Dad, how do trees grow?" Dad replied, "Do you really want to know?"

"Well trees need food like
me and you,
but they can make their own
which is a neat thing to do.

First, they take in some
rays from Mr. Sun
so for them daytime is
lots of fun."

"Then when it rains, and water comes from the sky, it enters their roots and travels up high,

to the leaves, which take in air, then Mrs. Tree mixes all these things with great care.

Out comes sugar which
is very sweet,
which is found in all the
fruits we eat."

"So plants make sugar as
their food,
which then put birds and
bees in a happy mood.

For plants like to share their food with others, just like you share your toys with your brother."

PHOTOSYNTHESIS

 **Sunlight** +  **Water** +  **Air**

=

Sugar

"Plants make food
through photosynthesis.
This big word you cannot miss.

So next time someone asks, "How do plants eat?" give them an apple that tastes so sweet,

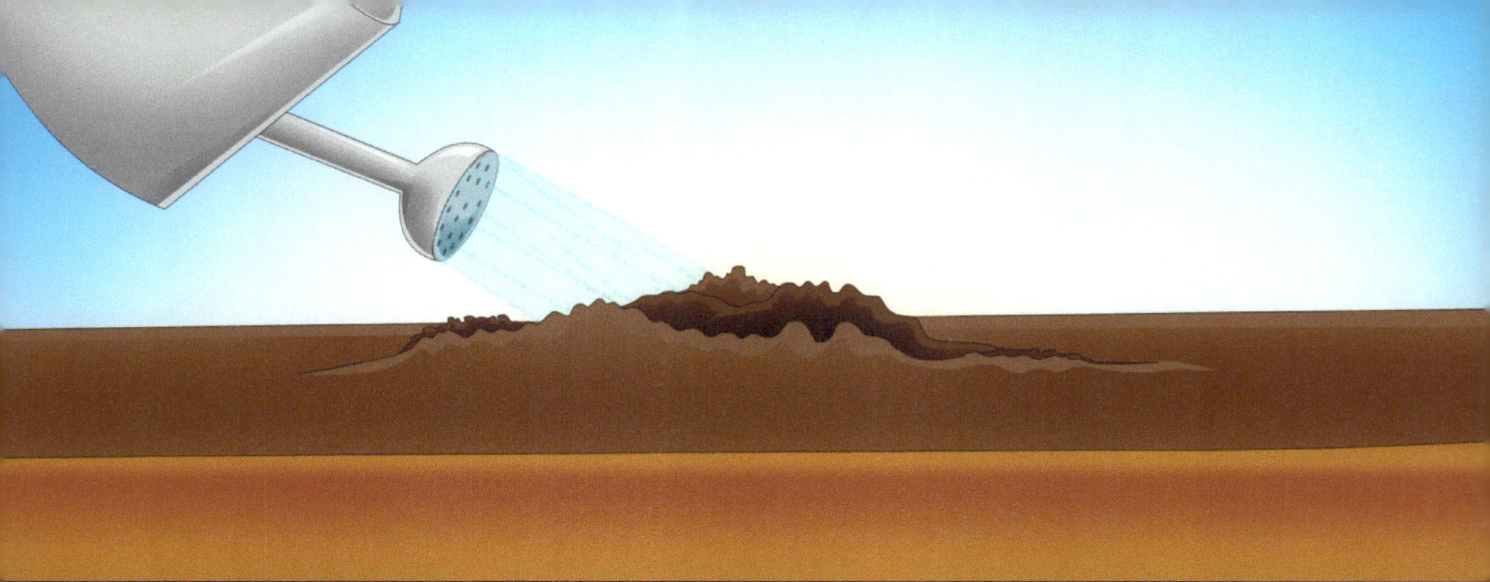

and tell them "Making sugar is the way plants make food to grow, through photosynthesis don't you know!"

Kelvin said, "Plants take in water, air and light and mix till it is just about right."

Dad said, "Right and that is the way trees get taller, so when you stand next to them you look smaller."

Kelvin said, "Thank you Dad. Now I know how plants are able to grow."

Explore Even More

Materials needed:

Plant seeds — You can use lima beans, sunflowers, marigolds, green beans or other types of seeds.
2 small pots (You can use cups as well), potting soil, and foil if you don't have a dark area to store your pot.

Young Scientists:

Here is an experiment that is easy to do:
It's about photosynthesis to give you a clue.
Place one seed in each pot filled with soil.
One pot should be placed in the sunshine, and the other pot should be placed in the dark, or covered with foil.
Water both seeds each and every day.
To see which one grows faster, and which one doesn't seem to change.
Did one plant grow fast or did one plant grow slow?
Carry out this experiment if you really want to know.

Young Scientists "If you seek an answer to a problem you can surely solve them."

About The Author

G. A. Sealy is from Queens, New York. He has a bachelor's degree in biology and education, and a master's degree in education and administration. G.A. Sealy is a certified health and wellness educator and consultant, life coach with an emphasis in weight loss, and has worked as a science educator in both New York and Georgia. His experiences as a science educator working in a variety of communities, led to his creation of "The Young Scientist Series" with the goal of "Teaching Young Minds through Science and Rhymes." He hopes that these books will inspire the young scientist in your child, and turn them onto the wonderful world of science, by transforming complex scientific concepts into fun, easy to read stories. Look out for other books in the Young Scientist Series, and the I love Me Series of books. For more information about the author, or any of his books, please contact dawitpublishing@gmail.com or visit www.dawitpublishing.com.

Dedicated to Tyler and Jayden true
Young Scientists.

www.ingramcontent.com/pod-product-compliance
Lightning Source LLC
Chambersburg PA
CBHW052049190326
41521CB00002BA/154